Patent Metallic Life Boats: Manufactured At The Novelty Iron Works

Francis' Metallic Life Boat Company

FRANCIS'

PATENT

ETALLIC LIFE BOATS,

MANUFACTURED

AT THE

NOVELTY IRON WORKS,

NEW-YORK.

New-York:
NTED BY J. P. PRALL, 9 SPRUCE-STREET.

1850.

Rescue of 201 persons with the **METALLIC LIFE CAR** from the wreck of the Ship Ayrshire, in a terrific snow storm 12th Jan., 1850, on the coast of New Jersey, by one of the Boats stationed on the coast by Government.—SEE PAGE 13.

CONTENTS.

REMARKS.

All new manufactures have the serious disadvantage that they conflict with existing interest. There is also an unwillingness to step out of a beaten track long followed, however prominent the merit of an improvement.

For these reasons the "Metallic Life Boat" was at first coldly received; all those however who then were doubtful of its capability, have become satisfied of its superiority to all others, and are now among the warmest patrons of the invention, and some of the most valuable testimonials now offered are from these persons.

The enormous loss of life by disasters to vessels of every description has called loudly for a boat capable of rescuing and preserving life from shipwreck, and experiments made for thirty-five years past has enabled us to obtain this desirable object. The substitution of metal for wood has made the Life Boat perfect.

The Metallic Life Boat is to be preferred to any other for the following reasons, viz:

1st. It is more "Economical"—no repairs being required, AND THE FIRST COST IS NO MORE THAN FOR A COMMON ONE OF WOOD, HAVING AN EQUAL AMOUNT OF AIR TO SUSTAIN PASSENGERS.

2d. Its unchangableness by heat of the sun or by fire—in either case there can be no shrinkage, but can be relied on as a safe retreat in any immergency. A great loss of life has occurred from time to time in consequence of the shrink age of wooden boats; such an instance has occurred withi the last week at the destruction of a barque, the heat of th sun having opened the seams of the boats to such a degre that they immediately filled with water, and twenty-fou lives were lost; and in case of fire the necessity of havin Metallic Life Boats is fully demonstrated by reference to th Steamer "Griffith," when the wooden boats were burned an thereby all hope of deliverance cut off.

3d. It is not liable to be broken or its safety impaired b any violence, such as striking against the vessel or rocks, a the testimonials will satisfactorily show. Indentations ma be made, and that to a considerable extent, but with a ham mer or even a stone they may be forced out without injur to the boat; whereas the same casualty occurring to a wood en boat would be attended, in most cases, with inevitabl destruction to all on board.

Very heavy expense has attended the perfection of th machinery and manufacture of the Metallic Life Boat; thi however is a matter in which the proprietors alone are con cerned, but, when an improvement is made for the preserva tion of life, should it not be duly appreciated by those whos safety is secured by the invention, and by every philanthro pist and passenger?

The Metallic Life Boat has been five years before th public; during this period its capabilities have been fully an satisfactorily tested by the most severe measures, and it i now considered by the Navy Department and Mercantil Marine as an indispensable appendage, and the only boat t be relied upon in time of danger. Reference may here b made to the Boats of the "Dead Sea Expedition," the Sloo of War "Albany," Steam Ships "Southerner," "Chero kee," "Crescent City," "Hermann," &c., &c.

The means of safety to passengers are various. Life has been saved by clinging to an oar, raft, float, or even pieces of wreck, when succor was near at hand, or within a moderate distance of land. In cases of Shipwreck on the Ocean or Lakes, it is obvious that the larger the boat and the greater its sustaining power the better, as a smaller one, or a raft, or oar would subject its occupant to almost constant immersion and suffering, and consequently weary out their efforts for self-preservation. In a good boat life may be preserved for days, perhaps weeks, and a distant land or friendly ship reached in safety. Deficiency in the capacity of boats has caused much sacrifice of life—all in the moment of danger rush INSTINCTIVELY TO THE BOAT, which, if small or not a Life Boat, is soon over-loaded, and that which should have been their means of preservation is from inadequacy and terror made the cause of their total destruction.

In cases like the burning of the Steam Boat "Griffith," where the Boats were all consumed and escape cut off, the advantage of a FIRE-PROOF METALLIC LIFE BOAT is strikingly evident.

The Metallic Boat is moreover lighter and stronger than any other, and cannot become nail-sick, worm-eaten, or water-soaked. Common iron was at first used for a few boats; none are now made but of COPPER or GALVANIZED IRON, and nearly eight hundred of this latter make are now in use, and many hundreds of lives have been saved from wrecks by them.

The boats now being furnished the U. S. Navy are made of copper and are calculated to last twenty years, at the expiration of which time they can be either repressed or the material sold for three-fourths of its original cost.

How few vessels built within the last fifty years are now in existence? And of these how very few have been broken up. Maritime records show that the greater part of these have been lost by casualties involving a very great loss human life. The statistics of such disasters uniformly sho that the heaviest loss of life has been among those term "crack ships or steamers," or in other words those of sup rior size and equipment, and deficient only in Life Boats.

Competition and custom has introduced into our merca tile marine all the comforts and luxuries of a home for t accommodation of the travelling public. Humanity and t preservation of human life also requires that the most a proved means of safety should be provided in case of acciden

Next to the ship, the Life Boat is the only means of safe to the crew and passengers, and the well-known durabilit and capacity of Francis' Metallic Life Boats as a means such safety, point them out as necessary, without whicl danger can be only partially averted in Fire and Wreck.

The following testimonials (unquestionable as to the veracity) are offered, that some facts relative to the "Meta lic Life Boats" may be more generally known.

TESTIMONIALS.

Awful Collision at Sea between Steamship Southerner and Bark Isaac Mead.

New York, October 5th 1850.

The steamship SOUTHERNER, which arrived here last night from Charleston, ran into the bark *Isaac Mead*, from this port, bound to Savannah, yesterday morning at 2 o'clock. The latter sank immediately, and *twenty-two* souls were lost! We give the account of this terrible disaster from the LOG-BOOK of the *Southerner:*

On Friday, at 2, A. M. lat. 38° 39', sounded in 22 fathoms water; relieved the wheel. In 10 minutes after we made a sail on the larboard bow: put the helm hard aport; stopped the engine and backed strong, when we came in contact; we backed clear and stopped the engine when the vessel went down under our bow, which was in less than five minutes from the time of the collision. Hearing the cries of distress in the sea, through the exertions of the crew and passengers, we were able to man three of Francis' Metallic Life Boats, and saved seven of the crew and two passengers out of thirty-three in all. She proved to be the bark Isaac Mead, from New York, for Savannah with a valuable cargo. She was steering SSW, we NNE; the wind to the North blowing strong, with a sharp sea; they unfortunately put their helm to starboard to cross us, as they saw us first, and took us for a vessel standing in shore. We remained until every vestige of her disappeared and nothing was heard but the moaning of the sea.

Too much cannot be said in favor of Francis's Metallic Life boats; had it not been for them we could not have SAVED ONE SOUL OF ALL ON BOARD; *a wooden boat would have been stove to pieces in lowering or coming alongside, the sea was*

*so bad.** To show with what facility they were got ready, in 4
minutes from the time the first boat was lowered and manne
by the second officer and two of the crew, she returned wit
seven; the second, manned by the first officer and two of th
crew, Capt Imbbock and Capt J. C. Berry, who nobly volun
teered their services; the third, manned by Thos. Vail and th
balance of the crew. When we gave up all hopes of findin
any more, we turned our attention to our own damage, foun
that we had carried away our cutwater, bobstay and flyin
jib boom, with the head rails and some scratches on the bow

These Metallic Boats were about the first made and hav
been used by the Southerner ever since she was built—
they are of small dimensions and imperfect constructio
compared with those of the improved kind made now. The
were used without floats or fenders, or any extraneous ai
presenting an entire metal surface alone inside and outsid
for that hard service.

From Capt. M. Berry.

STEAMSHIP SOUTHERNER,
New-York, Aug. 6, 1850.

SIR,—

When this ship was built, four years ago, I had her fitte
partly with ordinary wood boats and partly with Francis
Galvanized Iron Life Boats: the wood boats have long sinc
been given up, as they became *leaky*, *staved*, and *useless*, an
have been *replaced with metal;* while the Iron Boats remai
sound and useful and are at this time being cleaned for ne
painting, and are found to be *as good as new without eve
having had any repair* (though in that time they have bee
subjected to duty that would have destroyed a wood boat,
and have given me and my crew *unlimited confidence in the
in the worst positions conceivable.*†

* These are the same boats referred to by Capt. Berry, in his lette
of the 6th August, 1850, as having been in use ever since the shi
was built. They are long, narrow, and straight sheer, and small ca
pacity, and have metal air chambers only.

† These were the same Metallic Boats used so successfully at th
COLLISION AT SEA BETWEEN THE "SOUTHERNER" AND THE "ISAA
MEAD," OCT. 5th, 1850.

On one occasion of speaking another ship, in distress for ovisions, *her wood boats were too leaky to float, and we re obliged to use ours to supply her.*

They cannot either sink, burn, break or remain overset; ey are the only kind of boats in which unqualified confidence n be placed at all times, as they are always ready for use, d I would not now have wood boats at all. There is no mparison in the safety and durability, and above all in the portant confidence they give to the passengers and crew of ship in dangerous situations; they are the only boats that n be depended upon in case of a steamer or wreck taking e.

Signed, M. BERRY.

To Hon. John Davis, U. S. Senate,
Washington, D. C.

From John Maxen, Esq.

Squam Beach, Monmouth County,
New Jersey, March 13, 1850.

WALTER R. JONES, ESQ.,
President of the Board of Underwriters of New York.

R,—

I was present, and superintended and sent the line by the ortar on board the ship "Ayrshire," on the 12th of January, 1850, and by means of the Metallic Life Car, we landed safety her passengers, in all, TWO HUNDRED AND ONE, which, my opinion, at that time, could not have been otherwise ved, as the sea was so bad that *no open boat could have ved.* We attached the line to the shot and fired it from he mortar. It fell directly across the wreck, and was aught by the crew on board, and the hawser hauled off, to hich we attached the Metallic Life Car, and pulled her to nd from the wreck through *a terrific foaming surf.* EVERY OUL, MEN, WOMEN, CHILDREN, AND INFANTS, CAME THROUGH HE SURF DURING THAT COLD SNOW-STORM, DRY AND COMORTABLE. During the whole time of landing these persons, ne of the India Rubber floats put around the cars outside, y order of the government officer who superintended, was ull of water, and the other full of air, showing the ability of the Metallic Boat to do her work, even under such disa vantages as having air on one side, and the weight of th water in the India Rubber float on the other, in a heavy su The ship came on shore abreast of the station house, and th are ten miles apart; now if she had struck between tw houses, or even four miles from shelter, many of those w saved from drowning would have perished with cold, as was a *cold snow-storm at the time;* but, as it was, all we landed DRY AND COMFORTABLE, and no one suffered, as w immediately put them in the house, and found the benefit the fuel there provided by government, and this, in m opinion, shows the necessity of having the stations near together.

I have had much experience in wrecking, and was presen at the wreck of the ship "*John Minturn,*" and now say de cidedly, (and many others who were present at both wreck join with me) that if we could have had the mortar an Metallic Life Car, we could have saved a great proportion if not all of the souls from the John Minturn, which wa wrecked on this beach.

The car is also very valuable for landing *specie, jewels silks,* and *packages of all kinds,* that could not be saved b an open boat. We can also now communicate with the ship by means of the mortar and Car, as soon as she strikes, with out waiting as heretofore for the storm to cease, by which time she may go to pieces, and all be lost.

N.B.—With the above arrangements well attended, there need be few lives lost and much less property.

(Signed,) Yours, very respectfully,
JOHN MAXEN,
Wreck Master.

From Capt. Benj. Downing.

Extract of a Letter from Benj. Downing, Light House Keeper, Eaton's Neck, L. I.

September, 11, 1850.

I assure you that no person except myself and son, lef the shore in the "Government Metallic Life Surf Boat," an took a man off a wreck, named John Clark. The boat is 2 feet long, and rows six oars: I went alone with my son, be-

.use the *storm was so great that no one would go with me.* :X MEN STOOD ON THE SHORE AT THE TIME. I could not and on the beach, old as I was, and see a man perish hen I had the means to save him, and I have the reward my own breast. I am 66 years old, and my son 16.

Had I been present with this *Life* Boat, at the dreadful lamity of the bark Elizabeth, lost on Fire Island, last onth, I have the vanity to think no one need have been lost. I have been on the water all my life, and in all kinds of ssels, from a yawl boat to a seventy-four, and could not ve got off to the wreck in any other boat I ever saw.

From R. C. Holmes, Esq.

HE WRECK OF THE EUDORA.—THE METALLIC SURF BOAT.

Cape May Court House, Nov. 17, 1849—10 P. M.

IEUT. JOHN McGOWAN:

Dear Friend,—I am just off the beach, Ludlam's, immeately opposite the boat house, were there is a large steamer hore, the Eudora, from New York, bound to California. nowing your desire to know how the Metallic Surf Boats work, affords me great satisfaction to acquaint you. I landed all the ssengers this day and their baggage through a heavy northst surf without difficulty. My men remarked, "*It was ly fun to play in the breakers with her.*" She is the finest ing I have ever seen for the purpose for which she is tended and does the Inventor great credit. I have quainted Mr. W. R. Jones, of New York, with her perrmance.

I am, very truly yours, in haste,

(Signed) R. C. HOLMES,
Collector of the Customs.

New York, Dec. 3, 1849.

'alter R. Jones, Esq., President of the Board of Underwriters, New York.

DEAR SIR,—While landing the cargo of the steamer Eura ashore on Ludlam's Beach, Cape May, I had the honor address you, though hastily, and inform you, that the crew, passengers and their baggage, had been landed through surf with safety, in one of the Government Metallic Life S Boats, and under my care. Permit me, now that I have m leisure, the cargo, and passengers being again back New-York, to perform an agreeable duty, to the Gover ment, Mr. Francis, the manufacturer, and Lt. John McGow under whose care they were erected, and state more fully t performance of the Metallic Life Surf Boats, that they m be better known, more appreciated, and more extensiv adopted along our dangerous coast.

With these boats properly managed and manned, it a pears to me there can be little danger of loss of life shipwreck. They will live in almost any surf, and it m indeed be a terrific storm when a stranded vessel can n be boarded by them. If they are not entirely proof to t waves, nothing has ever been made to outlive them. Th are strong, light, lively, and are so constructed that they w carry their crews, when full of water. When kept head stern to the sea, they cannot be filled or swamped.

Our boatmen have so much confidence in them and consid them so entirely safe, that the difficulty of obtaining crews man them is no longer considered.

The surf must be indeed terrible when these boats whi we now have cannot go in safety.

In my district there are six Metallic Life Boats and s Life Cars, with Houses and Apparatus, all which t Government has furnished. They are about eight mil apart, and kept in constant readiness. And being of Met are always ready for instant use.

I have the honor to be, dear sir,

Truly, your obedient servant,

R. C. Holmes, Collector of Custom

Cape May, N. J.

Wreck of the sch. A. R. Taft, off Charleston, S. C.

Extract from the Log Book, May 6th, 1850.

As soon as information of her perilous position w known, several smacks, and a pilot boat went out her assistance, but the sea was so rough that th found it impossible to approach near enough to rescue t

:ew. Capt. Magee, of the steamer Nina, got his boat under 'ay at about 5, P. M.; and went to her aid, and having one f Francis Metallic Boats on board, launched her, and by his reat exertions took off the Captain and the crew and brought iem to the city between 8 and 9 o'clock last evening. Great redit is due to Captain Magee for his exertions. He proeeded of his own accord to the assistance of the distressed lariners, and had it not been for his promptitude, and the ict of having attached to his steamer the Metallic Life Boats ientioned, there is great doubt whether the crew of the A. R. 'aft could have been saved from a watery grave.

From Professor Grant,

Relative to the Copper Gig made for the United States Sloop of War Albany. DIMENSIONS—*Length* 30 *ft.*; *Beam* 4 *ft.* 4 *in.*; *Depth* 23 *in.*; *Thickness of Copper* 32 *oz.*; *Straight Sheer, like a Race-Boat.*

Ir. Joseph Francis, New-York.

SIR,—Although I have never had the pleasure of knowing ou personally, I know you well by reputation, and I deem ; my duty to lay before you the following facts with liberty o make use of them as you please.

Your exertions for so many years to bring to perfection an astrument to preserve human life from wrecks by Storm and 'ire, and in which you have been so successful, deserve all raise, and to withhold any information that might stimulate r aid you in the smallest degree in your efforts, would be nost criminal.

Washington, Sept. 30, 1850.

Being on duty in the U. S. service during the year 1848, n board the U. S. Steam Frigate "Mississippi," in the harbor f Sacrificio, near Vera Cruz, and requiring a boat to transort disinfecting material from the Castle of St. Juan de Jlloa, I made a requisition for one, and was informed that o boat could be had on the station for that purpose. I subequently learned, during the month of January, that one

of your Copper Life Boats was then lying a wreck in t
sand on the landing of the Castle. I immediately repair
to the Castle and discovered the boat lying in about thr
feet of water, half full of sand, and large pieces of old iron i
side some weighing 150 pounds. She was exposed to the bre
of the surf on the shore, and the wood work had be
broken, such as her seats, row-lock, and wash board, by a
parently heavy blows of a sledge, or a large iron bar lyi
near, as the marks of the same were plainly visible up
them. On cleaning out the stones, iron, &c., I discover
large concave indentations upon various parts of her side
made by heavy blows upon the *inside*, apparently with
heavy sledge, or bar of iron. These indentations were co
cave like a dish, but not broken through the copper. Th
were evidently made, however, *with the design of destroyi*
the boat, which was unsuccessful, from the yielding nature
the copper. On LIFTING UP the CEILING, I discovered fi
large holes in the bilge and bottom of the boat, apparent
made with the same instrument, which was more successf
in this part, as they had evidently been punched throu
while the boat rested on the coral rock, thus preventing t
copper from yielding or becoming indented. I repaired the
holes by PLACING A SLEDGE ON THE INSIDE, and BEATIN
BACK THE BURR FROM THE OUTSIDE with a ha
hammer, thus closing the fracture, apparently as strong a
tight, as when new. I then fastened the seats and launch
the boat, seemingly as good as new. The whole operati
certainly did not require an hour's labor, and in my opini
less time was employed in repairing the boat, than in the a
tempt to demolish her, if the marks of blows upon her we
any criterion.

The boat was 30 feet long, narrow, low, and straight, ev
dently not made for sea service, but for a man-of-war Gi
She *worked well, however, at sea*, as myself and two hands n
vigated her between Sacrificios and the Castle for sever
weeks, in *all kinds of weather*, and two or three miles of th
passage is in the open sea. On one occasion we were ove
taken by a severe Norther, which, by the time we reach
Washerwoman Shoal, within a mile of the harbor, was
powerful a gale of wind as to drive spray fifty feet high ov
the mole of Vera Cruz harbor; yet, in the face of thi

THREE of us brought this 30 feet boat safely to the Cumberland frigate under the lee of the Castle. I am confident any wooden boat of the same dimensions would have inevitably been lost under the same circumstances, as we were several times nearly half full of water by the sea breaking over. The buoyant air tanks kept the boat well up and we arrived safely on board.

The above boat formerly belonged to and was made in 1846 for the sloop of war "Albany," Capt. Breese, and had been thrown one side for what was supposed inefficiency, but she was proved to be the strongest, swiftest and safest boat in the Gulf Squadron, notwithstanding the unjust prejudice against her. She is still in use as a shore boat and good yet.

(Signed.) ROBERT GRANT.

Capt. Samuel L. Breese.

Extract of a letter from Capt. Samuel L. Breese, commanding U. S. Sloop of War Albany, August 16th, 1847.

In consequence of the leanness of the Albany aft, she sends so deep in a heavy sea or lying to, or becalmed, that she often dips up her stern boat full of water, which was the case with the Copper Gig. Not liking her I left her for the use of another vessel of the squadron. Barring accidents, however, I think a metal boat, in point of durability, is superior to a wooden one. This Gig had no gripes under the midships, when "dipped up full of water," and yet did not break down.

From Capt. H. Windle.

U. S. Mail Steam-ship Cherokee,
New York, August 12, 1850.

I have had Mr. Joseph Francis' Metallic Life Boats on board this ship from the time she commenced running. I take pleasure in adding to other statements of my experience, that I conceive them the most *effective boats* now in use, under all circumstances of danger and difficulty. That they will go comparatively in safety through perils that would destroy wood boats, has been proved to me by the fact that a part of my crew having taken one up to the head navigation in the Chagres River, and then set her adrift, came down and was restored to the ship AFTER THUMPI OVER ROCKS THAT WOULD HAVE BROKEN UP A WOOD BO before she could have passed through the channels in whi the rocks are situated, these boats have therefore my unq lified approval. (Signed)

H. WINDLE, Commander.

From Com. C. W. Skinner.

Navy Department, Bureau of Constructi Equipment & Repairs, July 1, 1848.

SIR,—I have the honor to present, through you, to t Minister of Marine of France, the model of a boat now buil ing of copper, by direction of this bureau; and if the B is approved, after trial, they will be adopted for the vess composing the United States navy. Inferior models ha been used in mercantile marine, and given satisfaction.

Two of the boats which recently descended the river J dan into the Dead Sea, were made of copper; one other wood *was dashed to pieces in descending the rapids.*

The officers conducting the enterprise state that the me boats alone were capable of resisting the violent concussio which they received in shooting over those descents, of whi they encountered twenty-seven; and the only alteration t copper boats sustained was from indentations by striki rocks, which, with a hammer, were readily removed. Th did not leak one drop; the air vessels at their extremiti rendered them very buoyant. They will wear a long tim and when no longer serviceable, the material goes far towar paying for a new boat.

Some have been built of galvanized iron, and are approv by the officers that used them. The one now building is the dimensions of a frigate's quarter boat, and as soon ready, will be put to a severe trial.

Those that were built for the exploration of the Dead S were made in Sections, for the convenience of transporti them over mountains on the backs of camels; they we disjointed or joined together with great ease at the place embarkation.

From the experience already had in the use of metal boats
y the mercantile marine, they are considered more economi-
al and superior in all cases to boats of wood.

I am, sir, respectfully, your obedient servant,

CHAS. W. SKINNER.

To M. Alex. Vattemare,
Washington City.

From Com. C. W. Skinner,

Navy Department, Bureau of Constr'n, Equip't., &c., July 19th, 1850.

SIR,—Having been informed that Congress was about
dopting measures for the prevention of the great loss of life
hich sometimes occurs from steamboat disasters on our
aters, I take the liberty of expressing my opinion, founded
n the various trials to which Metallic Boats, (furnished with
ir chambers,) have been exposed; that they afford the best
eans of safety to the lives of persons exposed, either by the
oundering of a vessel, or destruction by fire.

A cutter so fitted, 26 feet in length, furnished to United
tates frigate "Savanna," was reported to this Bureau by the
ommanding Officer, New York, to be capable of sustaining
side, from 25 to 30 men, when filled with water.*

The great advantage they possess over boats built of wood,
, that they are fire-proof, and are not liable to leak when
xposed to the action of the sun, *or be mashed when coming
n contact with a ship's side, or even rocks.*

Those used by Lieut. Lynch in descending the rapids of
ordan, although they received violent concussions in strik-
g rocks, *did not leak a drop*—the indentations thus pro-
uced, were removed by a common hand-hammer.

Had the unfortunate Griffith, on Lake Erie, been supplied
ith two of those boats, 30 feet long, and of suitable width,
any, if not all her passengers could have been saved.

Many vessels of the navy have been furnished with one for
e purpose of crossing dangerous bars, landing in a heavy

* See letter to A. Vattemare, dated July 1st, 1848.

surf, or lowering at sea in the event of a man falling ove[r]board: for such purposes I consider them superior to a[ll] others heretofore used in the Navy or mercantile marine.

Very respectfully, your obedient servant,

(Signed) CHA'S W. SKINNE[R]

Hon. Dan'l. S. Dickinson, U. S. Senator,
Washington City, D. C.

UNITED STATES SENATE, 30th CONGRESS.

Extracts from the Appendix to the Report of the Secretary of th[e] Navy to the Congress of the United States, being portions of Report made by Lieut. W. F. Lynch, U. S. N., of an Exami[-]nation of the DEAD SEA.

U. S. Store Ship Supply, November, 1848.

To the HON. J. Y. MASON,
Sec'y U. S. N. Washington, D. C.

SIR,—I have the honor to report, that, in obedience t[o] your order of Sept. 30th, 1847, I assumed the command o[f] this ship on the 2nd day of October following.

By your special order I obtained two Metallic Life Boats and while the ship was being equipped, procured the variou[s] articles deemed necessary for the successful result of th[e] contemplated attempt to descend the Jordan, and explor[e] the Dead Sea.

On the afternoon of March 31st, 1847, we succeeded i[n] landing the boats and all our effects. The next day w[e] transported them to the banks of the River Belus, near Acre where we pitched our tents that evening.

While encamped near Acre, we heard the most alarmin[g] accounts of the hostile spirit of the Arab tribes on each ban[k] of the Jordan,—these were, in a measure, confirmed by [a] party of American travellers, who had been attacked tw[o] nights previous under Mount Tabor,—but I had full confi[-]dence in our resources. Our route was over high mountai[n] ridges, and through deep and seemingly impassable gorges but we succeeded, and on the 8th launched the boats upo[n] the placid waters of the Lake Tiberias. For the purpose o[f] economy in the transportation of stores, I purchased a fram[e] boat, the only one of any kind on the lake.

On Monday, April 10th, at 2.30, P. M., we started. After eaving the lake, we passed the village of Semakh, on the eft, and soon came to the ruins of a bridge of the same ame. The fragments of the bridge entirely obstructed the hannel, except a narrow place towards the left bank, where he pent up water found an issue, and ran in a sluice among he scattering masses of stone. After reconnoitering the apid we shot down the sluice,—the leading boat was whirled gainst a rock, where she hung for a few moments, but was ot off without damage.

Tuesday, April 11th.—We started at 8.40, P. M., the curent at first two knots per hour, but increasing as we adanced, until we came to where the river, for more than 300 ards, was one foaming rapid, a number of fishing weirs, and he ruins of another ancient bridge obstructing the passage. After five hours' severe labor we got the boats through,—the Metallic ones without injury, but the frame one so battered nd strained that she sunk shortly after, and we were obliged o abandon her. *Had our other boats been of wood they would ave shared the same fate.* A blow that only indents a meallic boat, would fracture a wooden one.

Shortly after leaving the last ruined bridge, we descended cascade, at an angle of 60 degrees, at the rate of 12 knots, nd immediately after, down a shoal rapid, where we struck nd hung for a few moments upon a rock. We have to day lescended one cascade and seven rapids,—three large, and our small ones.

Wednesday, April 12th.—At day-break examined the vhirl-pool and rapids. Descended the first rapid, and at 1.05 shot through the whirlpool, and down a desperate ooking cascade of eleven feet. At 12 we stopped to rest, nd at 45′, P. M., started again,—and 1, P. M., completed the lescent of the third rapid since morning. Near sunset, we lescended the most frightful rapid we had yet encountered, nd, after passing down two others, arrived at 6.30, P. M., at Jir Mejami (a bridge of the place of meeting), and shooting hrough the main arch, descended about two hundred yards f the shallow rapid, when, it becoming too dark to proceed, ve hauled to the right bank and made fast for the night.

We to-day descended two cascades, and six rapids, four arge and two small ones. The current has been rarely less than four, and sometimes down the rapids as much as twelv knots per hour.

Thursday, April 13th.—Succeeded in getting the boat safely down the rapids, uninjured, save a few indentations i their bilge. As we would this day approach the lower Ghor which is traversed by hostile tribes, to be prepared, I mounte our heavy blunderbuss on the bow of the leading boat. A 10.40, *descending an ugly, brawling, shelving rapid, sh struck on a rock, just beneath the surface of the water, an broaching, broadside on, was thrown upon her bilge,* takin in a great deal of water,—but all hands jumping overboard her combined strength and buoyancy carried her safel over—though for some moments we feared she would go t pieces.

The river to-day varied from five to six knots velocity o current—we descended twelve rapids—three of them formidable ones.

Friday, April 14th.—The boats this morning had littl need of oars to propel them,—the current carrying then down at the rate of four knots per hour. The width of the river has varied from seventy yards, and two knots current to thirty yards, with six knots current. We struck thre times upon sunken rocks to-day,—and the last time, nearly lost the leading boat.

Saturday, 15th April.—We started at 8.24, A. M., and, at 9.34, passed an ugly rapid by Waddy Malakh, (ravine of salt) with a small stream of brackish water running down.

Monday, 17th April.—We started at 6.25, A. M., the river 40 yards wide and 7 feet deep, flowing at the rate of 6 knots, with much drift-wood in the stream,—many large trees, some of them recently uprooted. At 4.52, we passed down wild and dangerous rapids, sweeping by the base of a perpendicular hill.

Tuesday, April 18th, 1847.—At 3, A. M., we were roused with the intelligence that the pilgrims were coming,—and were obliged to move our tents a little higher up. In respect for the sanctity of the place, the boats were moored lower down on the opposite side,—but kept in readiness to rescue any of the pilgrims who might be in danger of drowning. At 1.45, P. M., started with the boats for Ain el Feshkah (fountain of the stride), on the north-west shore of the Dead Sea, a few ours distant. At 3.25 P. M., passed by the extreme point, here the Jordan is 180 yards wide, and 3 feet deep, and ntered the Dead Sea. As we rounded the point, a fresh orth-west wind was blowing, which soon freshened into a ale, and caused a heavy sea, in which the boats labored ccessively. The spray was painful to the eyes and skin, nd, evaporating as it fell, left incrustations of salt upon our ces, hands, and clothing.

Wednesday, April 19.—We noticed an entire absence of a-shells or of round pebbles upon the beach, which was overed with minute fragments of flint, and the foot-prints e made at landing, were, at our return an hour after, en-rusted with salt.

Made arrangements for camels to transport the boats in ctions across Jaffa, via Jerusalem.

It gratifies me to state that the boats are in almost as good condition as when we received them.

The above reference to the "flinty shore,"—the "foot-rint" so soon covered with salt,—the transport of the boats sections, show, that although hauled on such a beach every ight covered with sharp flint, and being constantly in such lt water as that sea, and roughly transported in sections, ere not destroyed, but returned in almost as good condition s when new.

From Joseph C. Thomas, Esq.

xtract of a Letter from Joseph C. Thomas, one of the Dead Sea Expedition, March 19th, 1849.

Being one of the party attached to the "Dead Sea Expe-ition," under Lieut. Lynch, I would state, that but for the etallic Life Boats, we never could have accomplished the escent of the River Jordan. *No other kind of boat could ave bounded over rocks, and down such deep and dangerous apids.* To these boats we owe our lives, for, had they failed, *s did the one wooden boat* we had, we must have been hrown on the shore and been murdered by the hostile Arabs, nd the whole expedition have failed.

From Capt. W. F. Lynch.

Extract of a Letter from W. F. Lynch, of the Dead Sea Expedi-tion, March 19th, 1849.

If in any way I can serve you by making known the ex-cellent qualities of your Metallic Life Boats, I feel bound t do so, FOR WITH NO OTHER KIND OF BOAT, HOWEVER STRONGL CONSTRUCTED, COULD THE DESCENT OF THE JORDAN HAV BEEN ACCOMPLISHED, and the expedition must have been un successful without them.

From Capt. J. Comstock.

New-York, Aug. 26th, 1850.

HON. JOHN DAVIS,

SIR,—My familiarity with numberless feats performed b Francis' Metallic Life Boats, and their perfect adaptation t be carried on ship board, warrant my saying that they ar superior to every other kind of boats in present use. I hop the Honorable committee of which you are a member will in their judgment, enact some wholesome law in regard t the general use of this invaluable boat, the advantages o which, over all others, are so great that I scarce know wher to commence enumerating them. I therefore will only sa that I am sure this is the best boat yet known for all lif saving purposes.

Very respectfully, your obd't servant,

(Signed) JOS. J. COMSTOCK,
U. S. Mail Steamer Baltic.

From Capt. J. Comstock.

New-York, Aug. 26, 1850.

SIR,—

I am fully of opinion that your Metallic Life Boats are in-valuable to all sea-going vessels. Their great strength an buoyancy renders them available when the ordinary woode boats would be of no service, and their lightness will allow o their being carried on ship board, where other boats could not be put. In case of fire no other boat of course is it ual, and on the score of humanity I hope all passenger-rrying vessels will be by law compelled to carry as many your boats as is consistent with room or space available for ch purposes. For the purpose of landing on rough beaches would be available when other boats would be dashed to oms.

Very truly yours,
Jos. J. Comstock,
Steamer Baltic.

To Mr. Francis,
Patentee of Metallic Life Boats.

From Capt. E. Crabtree.

xtract of a Letter from Capt. E. Crabtree, U. S. Mail Steam Ship Hermann, 19th June, 1848.

During the tremendous gale encountered by this Steam hip, 24th March, her two Metallic Quarter Boats on the lar-ard side were BLOWN OVER THE DAVITS SEVERAL TIMES fore we could get them in upon deck and secure them. ad the boats been of wood they must have been destroyed. I was present when the stern boat of the Steam Ship ashington came in contact with a post at the Novelty orks' dock.—She was twisted by the pressure at least two et and very much crushed. She was repaired at small ex-nse and now looks as well as ever. A wooden boat would ve been destroyed entirely under the same circumstances. find the Metal Boat is always tight and ready for use.

From Capt. E. Crabtree.

U. S. Mail Steam-ship Hermann,
New York, August 13, 1850.

Having used the Metallic Life Boats made by Mr. Joseph rancis, during the whole time since she was built, and ving added two others of larger size to my number, after aking several voyages I now have the pleasure to state, in dition to all former testimonials of my approval, that I still nsider these boats the most effective now in use, in all situ-ions of danger and difficulty; they are secure against fire; ey do not become leaky; they are always ready for use in

any exigency, and when they may be hastily and rough
used, are not injured by casualties that would either destr
wooden boats or render them useless when most needed, a
I unhesitatingly recommend their use generally.

C. CRABTREE, Commander.

From Capt. C. Stoddart.

Extract of a Letter from Capt. C. Stoddart, of the U. S. Ma Steam Ship Crescent City, 30th September, 1848.

I have received as good as it was the day I first receiv
it, the Galvanized Iron Life Boat which I sent you for r
pairs the evening of the 28th inst.—it having been crush
as it hung on the davits between the Steamer Crescent Ci
and another ship. Had she been of wood she would ha
been entirely ruined and beyond repair. She being repair
in six hours at an expense of five dollars and made as go
as new, I feel constrained to say that the Metal Boats a
far superior to wood, because they are always ready for us
are not affected by the heat of the sun or burning of a vesse
and are capable far beyond boats of wood of resisting th
action of the waves, and if jammed too, as mine was, can b
repaired at trivial expense, when a wooden one in like ci
cumstances would have to be replaced by a new one.

From Capt. G. W. Floyd.

New York, September, 13, 1850.

MR. JOSEPH FRANCIS,

DEAR SIR,—The Metallic Boats furnished by you for th
United States steam-ship Washington, when she was buil
are as good as new, and have needed no repairs. I consid
the Metallic boats superior in every respect, as they a
always tight and ready for use. In case of wreck or bur
ing of a vessel, a tight and fire-proof boat is of vital impor
ance, and at such times they are appreciated. From m
own experience, I can say that the Metal boats have th
requisites of safety and durability. Yours, truly,

GEORGE W. FLOY
Captain Steamer Washington.

From Capt. Thomas S. Budd.

Extract of a Letter from Capt. Thos. S. Budd, U. S. Mail Steam Ship Northerner, 22d August, 1848.

Francis' Patent Galvanized Iron Life Boats have given the satisfaction. They cannot become nail sick, worm eaten, or water soaked, nor leak, however much exposed to the sun, and are of course always ready for instant use in cases of emergency. As to economy, there is no expense to keep them in repair, and for preserving life they are always ready for instant use. For all the qualities for ship boats—such as durability, economy, capacity and safety, they are far superior to wooden boats.

From Edward K. Collins, Esq.

New-York, Aug. 4th, 1850.

Sir,—I have provided Francis' Metallic Life Boats for the "Collins' Line of Liverpool Steam Ships," as from my own experience they are far superior to any others in point of *Economy, Durability and Safety.*

I should think no other kind of boat could be relied on for spare boats for passengers when Steam Boats are destroyed by fire. They are fire-proof and are not affected by the heat.

Yours respectfully,
Edward K. Collins.

To Hon. J. P. Phœnix,
Committee Commerce, Washington, D. C.

From Capt. Thomas Brownell.

Extract of a Letter from Capt. Thos. Brownell, U. S. Navy, 11th September, 1848.

These boats are always ready for use at a moment's warning, in any climate, and such as would render a wooden boat entirely useless. They can be used at sea and lives saved where wooden boats could not live even if in a floating condition. The boats I have in use I have put to very severe tests.

From Walter R. Jones, Esq.

Extract of a Letter from Walter R. Jones, Esq., President of the Board of Underwriters, to Hon. James G. King. June 26th, 1850.

I am decidedly in favor of Galvanized Iron Boats, the air chambers can be sufficient in number to float all the passengers that can get in and around her sides.

The controlling advantages of the Metal Boat are, THAT THEY DO NOT BURN, and can always be kept tight and fit for immediate use, whereas boats built of wood are sure to leak like riddles, as they generally are placed where they are exposed to the sun and rain.

From Report of the Judges on Naval Architecture.

Extract from the Report of the Judges on Naval Architecture, appointed by the American Institute, at the 20th Annual Fair. Judges: Jacob A. Westervelt, Ship Builder; Thos. Brownell, U. S. N.; John H. Rhodes, Naval Constructor.

The Galvanized Iron Life Boats have superior advantages over all others for the following reasons:

1st.—Their endurance under severe trial, it being almost impossible to meet with sufficient injury to disable them from sustaining their compliment of persons for any length of time, in case of STORM, WRECK, OF FIRE.

2d.—Their extreme lightness, united with great strength.

3d.—Their inability to become nail sick, worm eaten, or leaky from exposure to the sun, however long they may be out of water.

4th.—These Life Boats may be used at sea to preserve life when nothing else can live, or for the daily use of the ship, *being always in readiness for either service.*

From Capt. M. Berry.

STEAMSHIP SOUTHERNER,
New-York, *Oct.* 23, 1850.

MR. J. FRANCIS,

SIR,—I send you an order for a suit of Metallic Life Boats for the New Steamship "General Marion," now building for the Charleston Line. They will be required in January. Yours, &c.,

M. BERRY.

From Capt. Tho's S. Budd.

New-York, October 21*st*, 1850.

Ir. J. Francis,

Sir,—I herewith send you an order for a suit of Metallic ife Boats for the new Steamship building to take the place f the Northerner.

Those ordered by me for the Northerner when she was uilt are yet as good as ever, having been in service ever ince without repairs—I therefore renew my orders for more.

Respectfully yours,
Tho's S. Budd.

From E. Mills, Esq.

New-York, October 19, 1850.

. Francis, Esq.

Enclosed please find an order for your Metallic Life Boats, ne set each, for the new Steamers "Louisiana" and "Mexi-o," building for Messrs. Harris and Morgan, Gulf Line; nd one set for the new Steamer "Brother Jonathan," uilding at Williamsburg, for the Pacific.

The Metallic Boats ordered by me for the Steamships Washington" and "Hermann" when they were built, are et as good as ever, and have required no repairs.

I am satisfied, from my own experience, that the Metallic oat is far superior to any other in every respect.

Yours truly,
Edward Mills.

From Lieut. Washington A. Bartlett.

National Hotel,
Washington, Dec. 14th, 1850.

SIR,—

In answer to your inquiries as to my observations on the character and performances of your Metallic Boats on the coasts of California and Oregon, and of their general value as compared with wooden boats of like capacity, &c., &c., I have to state that,

There is a very large number of your boats in use in California and Oregon, in every possible variety of employment to which a boat can be put, from the largest to the very smallest; and of their ability to endure hard service beyond that of a common boat, no one could doubt, after seeing the rough handling they get there. In no instance have I seen one out of repair.

The two that I purchased of you have done all they were expected to perform, and both were so much in favor with others, that they were often stolen from their moorings, and for this reason they were sold to avoid the annoyance of seeking for them; they are still in use in the Bay of San Francisco.

At the mouth of the Columbia the Pilots have one of your small size, which they keep as a *safety boat*—and in one of the same size, Major Hathaway, U. S. A., with a party of seven persons, crossed the North Breakers of the Columbia River Bar—accidentally going to sea in a dark night, and recrossed the Breakers the next day, *without shipping water*. This was a feat for a *landsman* to perform, of no ordinary character!!

The large boat which you furnished to my friends who went round the Cape, left the ship with a party of nine persons, when seventy miles distant from the Island of Juan Fernandez, and in a gale which, the same night, reduced the canvass of the ship to a close reefed top-sail, causing those on board the ship to despair of ever seeing them again, the little Metallic Boat rode it out in safety, and the whole party safely landed the next morning, while it blowed so hard the ship could not approach the shore. I have used your boats with great satisfaction.

I know that in the Tropics, where there is a great shrinkage, and where boats of wood so soon become "nail-sick," and where the worm is so destructive to wood, *your boats are always ready for service*, without the necessity of repair, while they will endure without injury an amount of thumping on rocks and beaches which will destroy an ordinary boat.

The Boat you supplied to the U. S. Ship Vincennes, Capt. Hudson, was in daily use, and, for lightness, speed and safety, the favorite boat of the ship. The 1st Lieutenant stated to me that her performance was admirable—that she did all the work of the ship. Such a boat, where an economy of men is desirable, is of the first importance in the equipment of a ship.

Having some years since seen your first essay in the construction of the Metallic Boat, both for ordinary use, and for Life Boats, I have, with no little interest, watched the progress you have made in overcoming prejudice and opposition, and consequently, wherever I have been, when meeting with "Francis' Metallic Boats," I have been particular to inquire into their performance, more especially among seamen; and I have found in the last two years, whether from actual trial, or the widely disseminated testimony of intelligent seamen, who have fully proved their merits, that there is now little opposition to them, in any quarter, their superiority for durability and safety being generally admitted.

For my own part, in fitting out a vessel for any service, I would not fail to supply her with your Metallic Fire Proof Boats, both for safety and economy.

Yours, very truly,

(Signed,) Washington A. Bartlett,
Lieut. U. S. Navy,
Ass't Coast Survey.

To Joseph Francis, Esq.,
Washington.

Eagle. Eagle.

From Report of Secretary of Treasury.

Extract from the Annual Report of the Secretary of the 'reasury to Congress, 1850.

"Measures have been taken promptly to execute the design Congress in providing for the security of life and property the sea coast. Metallic life boats with the usual fixtures, signed for five points on the coast of Florida, and three for e coast of Texas, have been contracted for. Like facilities, th the addition of mortars, shot rockets, and station houses, ve been authorized along the shores of Long Island, cluding a station at Watch Hill, in Rhode Island."

The Metallic Boats referred to are to be similar to those w on the coast of New Jersey and Long Island, provided r by Congress under an appropriation in 1848.

CPSIA information can be obtained
at www.ICGtesting.com
Printed in the USA
LVIC04n1432190713
343749LV00003B
9781120016966